I0060609

MATH IS BAD

A brief look at the other side of the coin
Kaveh Mozafari

ExcellenSation Inc.

370 Steeles Avenue West, Thornhill, Ontario, Canada, L4J6X1

Visit the author's website at www.kavehmozafari.com

Printed in Canada

First Edition 2016

ISBN: 978-0-9940739-3-8

Contents

Dedication

I would like to dedicate this book to a very dear person in my life, whom I haven't had a chance to talk to or see for years. He is a great nobleman with a great personality, and he was kind enough to choose my name at my birth. I have always been fascinated by the amount of knowledge he carried, his great insight and generosity. Dr. Manoucher Bigdeli is an inseparable part of my life, and I hope dedication of this small text shows my great respect for him.

Preface

Mathematics... words cannot describe its dedication to humanity. It is a trusted friend, which served us well. Its complication made us think. Some of us ran away, and a few stayed and stared right into it. They enjoyed and fought. Some returned from the joyful battle and described the beauty of the beast. A few explained the monster to be manmade, and others recognized it as the reflection of nature.

Regardless of how we ended up with this charismatic, friendly, loveable creature, we take advantage of its presence. However, as we advance, we notice the absolute sense that this charismatic friend conveys is possibly not the best picture we need to look at to learn more about ourselves and the surroundings.

The mentioned doubtfulness strengthened a branch of mathematics known as statistics. Statistics opened new doors to a better understanding of many phenomena. Disciplines such as psychology and physics experienced incredible revolutions as a result of the statistical tools they acquired.

Moreover, the fact that statistics does not have a firm opinion about many phenomena makes it more realistic and approachable. Most importantly, statistics consistently reminds us we know nothing for sure, and anything is possible.

I purely believe we should bow down to the complexity of the cosmos and respect it. We should try to accept we can only get close to reality with an estimation. The reason for that is approximation seems to serve as the building blocks of the world. Consequently, statistics is one of the best ways to familiarize ourselves with the reality.

Acknowledgments

First and foremost, I have to acknowledge the unconditional support of my parents and my two sisters, Sahar and Sara. They back me up for whatever it is that I believe in, even when it may not make sense to the rest of the world.

I would also like to thank my friends who accepted me unconditionally. It is a great feeling that they still invite me to their gatherings even though I miss most of the events. I must thank whoever I had interaction with as I am sure they did affect me and my little works in some ways.

Introduction

"Math is Bad" tries to reveal the other side of mathematics by showing the uncertainty in every phenomenon in the universe. This book seeks to demonstrate that the absoluteness is not absolute, and we always have to leave room for any possibility.

The flow of the book is set up in a way to initially explain the history of mathematics and how our ancestors used math. Then the book focuses on the relationship between logic and mathematics. In doing so, the book tries to close the gap between reality and logic. Afterwards, the book attempts to show a small pinch of mathematical beauty by providing a few examples.

The following chapter attempts to demonstrate that mathematics is too simple for the multifaceted world. On the same note, the book exposes statistics as a useful tool in real life by providing some interesting examples.

During the preparation of the book, we tried to avoid complexity while explaining the claims and arguments. A simple tone and avoidance of detailed mathematical or scientific findings are kept in mind during the preparation of the book. As a result, this book is understandable for any interested audience with any background.

Birth of Mathematics

Many good sources discuss the history of mathematics, specifically the beginning of it. Some pieces of evidence reveal interesting links towards the origin of mathematics. For example, the Ishango bone that was found in Africa suggests maybe people over 20,000 years ago had some ideas about multiplication, division and, possibly, prime numbers. Prime numbers are whole numbers (integers) greater than one that are only divisible by themselves and one. Doesn't that sound shocking?

Assuming that numbers have some building blocks, we can say every number is either originated from the multiplication of prime numbers or is a prime number. For example, 7 is a prime number, and 10 is originated from the multiplication of two prime numbers; namely, 2 and 5. Now, the question is, does it mean we can look at a number only through multiplication? That is, why don't we say 10 is the addition of 9 and 1 ($9 + 1 = 10$) instead of multiplication of 2 and 5 ($2 \times 5 = 10$)?

The answer can be that multiplications are additions in nature but at a faster rate. For example, how many moons do you see in the following figure?

Figure 1

Correct, thirteen. I assume you counted them one by one. You added them up one by one. Now I would like to ask you how many moons you see in the next figure.

Figure 2

Did you say 30? Good job. Which one was easier to count? You could count them one by one, or you could have used the grouping that is provided to multiply six by five since we have six groups and each group contains five moons. Although the number of moons in the second figure is more, it is easier to count. The reason is that, in the second one, we could use multiplication to determine the total.

You may argue we do not have so many moons, you are correct; we do not. However, someone might. As I write this chapter (May 15, 2016), scientists have counted 67 moons that are entertaining Jupiter. One of the moons of Jupiter, Europa, may even have water.

Now, getting back to the bone from 20,000 years ago, did people back then know about multiplication? We cannot answer solely based on the prime numbers found on the bone.

There are some insects, called cicadas, which use prime numbers as a timing scheme to come out of the ground of the eastern United States forest. The cicadas do so every 13 or 17 years. It is highly unlikely that the cicadas know the definition of the prime number.

What I am trying to say is that mathematics revealed itself as a reflection of the universe before becoming an impressive set of rules and definitions. Therefore, the insects are not necessarily aware of arithmetic to get out of the soil every prime number of years. In turn, their behaviour is a result of survival advantages that they have received from this strategy. It seems they have mastered timing through random trial and error.

The cicada's strategy makes it harder for predators to predict their exact merging time. Cicadas are fascinating creatures, and I highly encourage you to investigate them further, particularly the cicada's signaling and behavioural theories.

Now, the question is, did we know about the prime numbers' properties 20,000 years ago? How do we identify what we knew for sure regarding mathematics years ago? Some concrete pieces of evidence confirm we recognised a lot of mathematics around 1900 B.C.

For example, from Babylonian clay (Plimpton 322) dated 1900 B.C., we believe people back then had knowledge about Pythagorean triples. Pythagorean triples are whole numbers (integers) that can create the sides of a right triangle. Therefore, they must follow the $a^2 + b^2 = c^2$ equation, for example, $3^2 + 4^2 = 9 + 16 = 25 = 5^2$ or $3^2 + 4^2 = 5^2$. Hence, 3, 4 and 5 are among Pythagorean triples. It worth saying Pythagoras was alive from 570 B.C. to 495 B.C. Even though the clay is almost half a millennium older than Pythagoras, it is named after Pythagoras. I believe it is fair to call Pythagorean triples, Babylonian Triples, based on the presented evidence.

The other acceptable pieces of information that we can relate to the birth of mathematics are some papyri found in Egypt. In particular, we can discuss one papyrus, Rhind Mathematical Papyrus (RMP). RMP dates back to 1650 B.C. and contains a few reference tables along with 84 questions.

The papyrus contains three books that include problems in various mathematical disciplines, including algebra, arithmetic, mensuration (issues dealing with volume and slope of pyramids), fractions multiplication, mathematical application (finding the strength of bread and beer) and geometric progression.

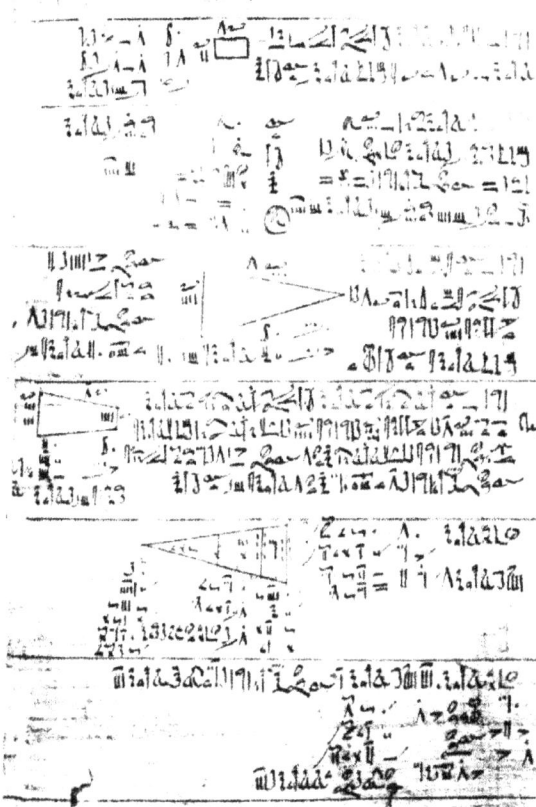

Figure 3. The Rhind Mathematical Papyrus (RMP) (Papyrus British Museum 10057, or pBM 10058).

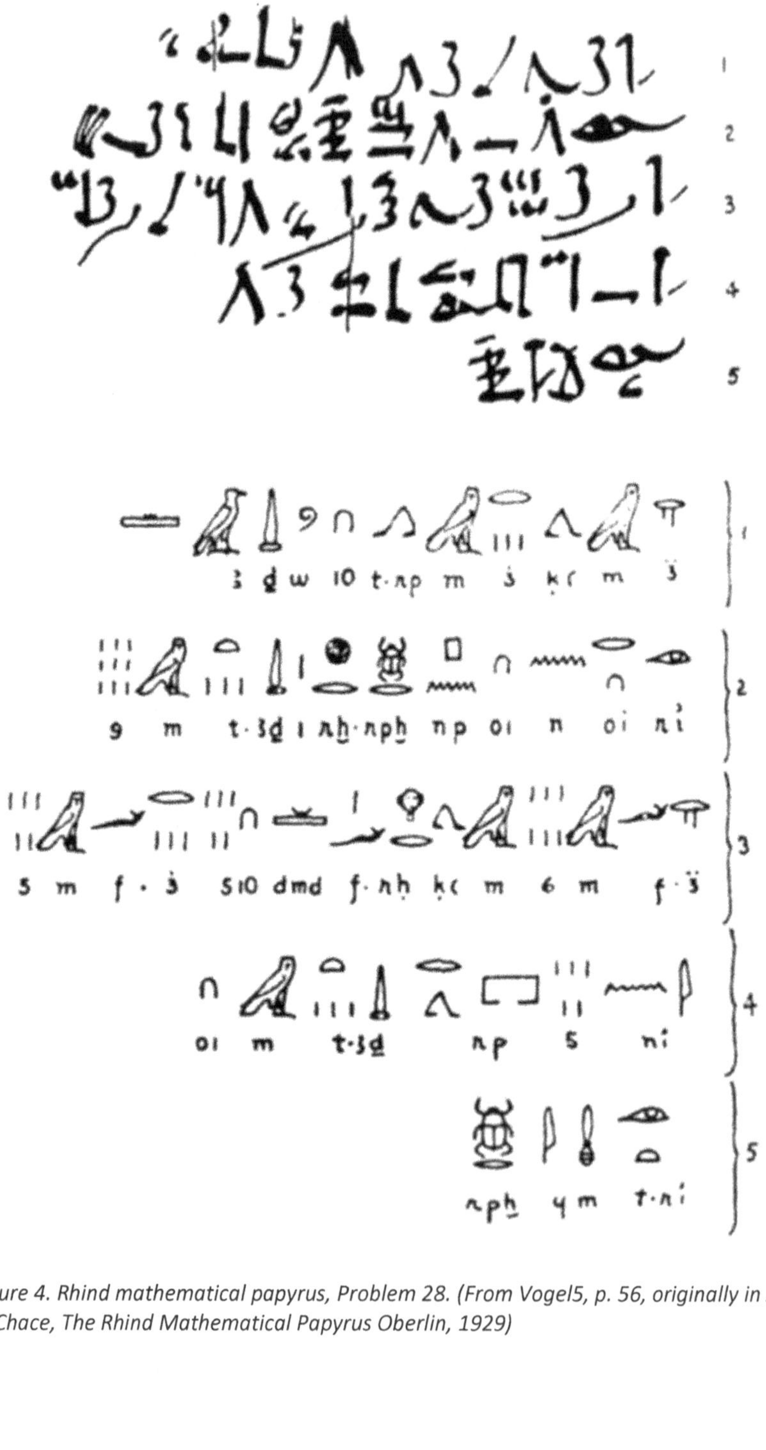

Figure 4. Rhind mathematical papyrus, Problem 28. (From Vogel5, p. 56, originally in A. B. Chace, The Rhind Mathematical Papyrus Oberlin, 1929)

The obvious point is that people during the early period needed mathematics to build, trade and cook properly. The aforementioned books (Rhind Mathematical Papyrus), along with another papyrus, are great examples of why they needed mathematics in the first place. Please note that the problems addressed in the papyri are, by no means, simple. For example, when we talk about volume calculation, in the era we live in, we start learning how to calculate the volume from grade 9. We develop our knowledge to find out ways to calculate the volume of the most shapes with more precision in the second and third part of university calculus (mostly in the first two years of university).

On the other hand, as a person who studied with high school and college students, I can tell one of the most problematic yet seemingly simple parts of math is working with fractions.

We can see in the papyri we talked about, the author addressed the fraction issue by providing many examples and a reference table.

The tables are rather remarkable. One tells the readers how to write $\frac{2}{n}$ (n is an odd number between 3 and 101 inclusive) as the sum of, at most, four unit fractions. By unit fractions, we try to emphasize the numerator must be 1. Here are a few examples.

$$\frac{2}{3} = \frac{1}{3} + \frac{1}{3}$$

$$\frac{2}{15} = \frac{1}{10} + \frac{1}{30}$$

$$\frac{2}{101} = \frac{1}{101} + \frac{1}{202} + \frac{1}{303} + \frac{1}{606}$$

Fractions were problems back then too. Even to this day, we put a lot of effort into learning them, not because it feels good to see people struggle when they cannot work with fractions, but because there are many practical usages for them. Is it creepy to say I may feel good seeing people struggle with fractions? I didn't say it.

Speaking of possible usages of fractions, let's see what happens if you don't have money to buy a whole box of flowers that I am selling. What if you cannot work for a full day? Now this obsession with fractions leads to something interesting. How much can you divide and still get a measurable piece?

The smallest meaningful piece is a topic that we will discuss in more detail in another chapter, but we can elaborate a bit more on this. Briefly, you can divide only so much. That is, the concept of quantum makes sense only when we put a limit on our division as we need the smallest compartment. I will present my reasoning in the proper chapter.

I assume math also started to attract some people for other purposes. That is, only because it feels right to struggle, think, solve, doubt or predict; alternatively, in a more mathematical sense, to realize the limit of our knowledge approaches zero.

For some people who enjoy solving problems, it is a voice that something greater than daily life essence calls them. Something that enforces math lovers to do math not only to calculate the strength of beer but to know how accurate they can measure the area of a circle. In other words, how we can gauge the value of pi (π) to billions of significant figures (digits).

Maybe you don't care, maybe I don't care, but some people need to know it. Why? They enjoy the process. I call it love of math. It is a deadly disease, and it is contagious. Why deadly? The same reason that whoever drinks water would die.

The birth of math, for whatever reason, was elevated to something else. It helped people to think outside their comfort zone. That is, even though 20 silver coins is greater than five gold coins, the silver ones are worth less. Why? Simply because gold is more scarce in general. What did I say? "Why?"

Let's now move to the next chapter, and, in there, we are hopeful to find answers to a few of the "why" questions.

Mathematics Is Logical

You have two goats. They decided to date and eventually got married. After the ceremony and the talking part, they have four kids. Therefore, they become a family of six. Did you follow what I said? Well, it was mathematics, logic or a simple story that tangles a few facts together in an understandable manner.

What about this story? I looked at a girl, and she looked back at me. I decided to go to her desk (not table) and ask her out. I did as I said.

- She told me: "Are you crazy?"

- I replied: "Why?"

- She continued: "Because it was just a stare."

- Then I replied: "You are entirely correct, sorry to bother you."

I then proceeded to the other side of the room. I don't know if she ever looked back at me again because I didn't check.

I went home and noticed she didn't exactly say, "I won't go out with you," at least in a direct sense. Maybe she asked about my craziness to make sure I am a good candidate to go out with her. Moreover, I notice my answer to her question was blank. I did not answer her that I am not crazy, even though my belief is that everyone, including myself, is insane to some degree (fine, I am a bit more insane than others). The good thing is I did not lie. Wait, I said you are entirely correct, which meant I did confirm I am crazy. However, is her perception of craziness the same as mine? If it were, she wouldn't ask it in the first place. On the other hand, maybe if we have different perceptions, we should not have gone out in the first place.

Do you feel the daily confusion a person can experience? Believe it or not, there are more confusing materials and context to think and discuss over the course of one's life. It may be fair to assume these misperceptions can become frustrating, so people need to associate them with relaxing factors.

As humans could quantify and explain many things with numbers and simple relationships, they felt less of a need to associate everything with the relaxing elements. Also, I believe those who had real faith in a superpower could gain more firm evidence as they could associate things more logically with the superpower. In other words, dependent on the viewer's internal belief, individuals may interpret a similar phenomenon differently. Hence, the same events that strengthen someone's faith in the superpower existence may diminish someone else's belief in the superpower. I do not intend to dive into the superpower-related topic at the moment as it requires a separate book to present my claim.

As we were explaining, math is a lot easier to follow. The fact that mathematics is stepwise to a great extent led mathematics to be a great language to track and be "logical." What I want to get into is logic emerged from mathematics. In better terms, logic and mathematics are from the same parents, while mathematics opened a lot of doors to thinkers.

I get sick when I don't wear enough clothing during the winter. Therefore, I have to wear some warm clothes if I don't want to get sick. Do you see similarities between the goat problems I stated earlier in the chapter and the last few sentences? I have two goats, and if they bring four more, then I will have six goats. We are trying to put emphasis on the "if" and "then" parts of the statements. If we have two apples, and I eat one, then there is one left for you to eat.

$2 - 1 = 1$ Well, these facts seem interesting, at least to me. We will attack these problems in future chapters. The main point is that logic that we talked about is very easy to follow. Therefore, it is understandable. Now the question can be, is it applicable to everything? Well, the answer to this problem is rather hard, and we would try to counter it in the following sections. However, as a quick response, I would say, no, it is not applicable to everything, and you will see my reasoning very soon.

There are various types of logic. We believe it is a good idea to introduce and discuss their association in a very concise manner.

One of the most widely used logics in calculus and artificial intelligence is syllogistic logic. Aristotle (384 BC) put enormous effort on syllogistic logic. The Organon is the fascinating body of his work on logic. There are six set of tools that Aristotle employs in logic. In prior analytics, he introduced the syllogistic method. We have three kinds of syllogistic conditions. For example, If I go over the speed limit in front of an officer, I will get a ticket. I will not go over the speed limit in front of an officer. Therefore, I am not getting a ticket. I am hoping that you have a lot of questions and doubt the validity of the aforementioned logic, the reason being we need doubtfulness to prove many points later on.

That is, you may argue, I did not go over the speed limit in front of an officer, but I did get a ticket since I hit a pole. The officer used the amount of bend on the pole, along with the mass of the car, to calculate my speed. Therefore, I did receive a speeding ticket after all. You may come up with more similar arguments to say the initial claim is not necessarily accurate all the time. Now let's look at the following.

If $x = 2$, and $y = 3x$ then $y = 3 \times 2 = 6$

What I am hoping to get to is that the logical steps taken in mathematics are very similar to the path taken in the conditional syllogism. Please note there is no ambiguity in mathematics as opposed to the logical sentence that I used, which was ambiguous to some extent.

You may argue that I used a bad example, but I used a real-world example. The whole point of this book to take advantage of the real-world problems rather than isolated cases. We discuss the realistic point of view of this text in more detail as we proceed.

The other kind of syllogism is disjunction syllogism. Here is an example of it. The light is either on or off. The light is not on; therefore, it is off. In the mathematical form, "x" is either "zero" or "not zero." If x is not "not zero," then it must be "zero." Beautiful! We humans have used this type of logic in many applications, including building the binary language of computers.

Logicians have defined the disjunction syllogism system for only two states. Nonetheless, the question is, couldn't it be more than two states? At least for computers it can, when we talk about the new generation of computers which use quantum computing. Hence, having two might be useful and serve certain points, yet to prove there are only two possible states requires more rigorous work. For example, in the provided quantum computing example, we noticed that we can employ more than two critically beneficial states.

The other form of syllogism logic is the categorical syllogism. Let me give you an example to briefly explain categorical syllogism. All birds have wings. Ducks are a kind of bird, so ducks have wings. We can formulate the presented model using mathematical language in the following form:

$$N \in R, x = 2 \text{ and } x \in N => x \in R.$$

N Refers to "Set of Natural Numbers"

$\in$ States as "belongs"

R Refers to "Set of all Real Numbers"

$=>$ Reads as "therefore"

What I just said is that all natural numbers are real numbers. Variable x is equal to 2, and it is a natural number. Therefore, x is a real value.

As before, there is no confusion in the mathematical term. Mathematics talks with certainty and leaves no room for doubt. It speaks with confidence, and you can only be a listener. Two is a real number, period. However, in the presented example, for categorical syllogism, you may doubt the existence of any bird without wings. Well, the solution to this doubtfulness is you have to check every single bird to conclude wingless birds do not exist. The reason for that is, unlike numbers that are well defined by humans, we cannot identify all birds unless we have information about every single one of them.

With regard to the possible existence of a wingless bird, moa is the name of the wingless bird, which used to live on Earth. Although moa went extinct, it serves the point. In other words, since the nature that we try to apply logic to is not manmade, we do not have information about every creature in it. Later on, in the presented book, we get back to these types of contradictions in more detail.

The other form of logic is known as propositional logic, or propositional calculus. The naming gives us enough clue about the amount of mathematics embedded into propositional calculus. The roots of propositional go back to Chrysippus (279 BC), and the main difference between propositional and syllogistic is the use of propositions versus terms.

In the propositional logic, we have propositional constant. For example, the car has gas in it. The propositional constant can be either true or false. In the above example, the car either has gas or does not have gas. The cited statement is called a truth function. If I show the truth (the car has gas) with P, then $\neg P$ (read not P) means the car does not have gas.

Now, for any pair of propositional constant, there are four possibilities. Given that P and Q are two propositional constants, the following results:

P is true, and Q is true

P is true, and Q is false

P is false, and Q is true

P is false, and Q is false

The first relationship is a conjunction as both propositional constants are true (and), disjunction occurs when only either P or Q is true and usually signifies with an "or." For example, you are sitting or reading. The mentioned term has any of the following meanings:

You are sitting and reading

You are sitting but not reading

You are not sitting but reading

However, you could not be standing and not reading if I were truthful. The material condition connects the two propositional constants, which we can demonstrate as $P \rightarrow Q$ (If P, then Q). For instance, if I study, I will pass the course. I could have made it biconditional, which we show as $P \leftrightarrow Q$. We can read the mentioned as P happens if and only Q happens. We can also interpret this by stating if P, then Q and if Q, then P. Please note both conditions must be met. Let's provide an example. If you love your spouse, then your spouse loves you, and if your spouse loves you then, you love your spouse.

I don't think I need to discuss the relationship between propositional calculus (logic) and mathematics. However, I have a question about the propositional constant. What if the propositional constant is not a constant? Well, believe it or not, it is the case in many situations in our surroundings.

How can you guarantee propositional constant is not changing? Of course, we can think of more questions that we need to address. As a sample, another question can be, is $P \leftrightarrow Q$ equivalent to $Q \leftrightarrow P$? Yes, they are equivalent. What if you consider the ordering and timing? You may argue we can use special relativity to say it does not matter which occurs first as we can change the observer and have the order reversed. I am assuming we are not changing the observer, and the frame of reference is fixed. Having a fixed frame of reference is an issue that needs to be considered with more care in another context.

Well, is there any perfectly constant frame of reference with no acceleration? I don't think so as the whole universe is expanding. Even on a smaller scale, the earth is on an elliptical path around the sun, and our sun revolves around the centre of the Milky Way galaxy. It may be interesting to know that it takes our solar system roughly 250 million years (1 galactic year) to revolve around the centre of the Milky Way galaxy. The speed of our sun and, therefore, the solar system in the galaxy is close to 828,000 km/h, or 514,000 mph. We will get back to this discussion in further chapters.

The other kind of logic is known as the predicate logic, which can have several meanings. In the most basic and general terms, we can say predicate logic is a statement that can be true or false depending on the value of the variables.

We can explain predicate logic in more sophisticated terms than any other kind of logic. It is a system to clarify the first-order, second-order, many-sorted and infinity logic. We will discuss a few of these logics as we plan to go along in a concise manner, yet you are more than welcome to read more about them.

The first-order logic is also known as the first-order predicate calculus, the lower predicate calculus, and the quantification theory. The naming reveals enough association between the first order logic and mathematics. However, we can still propose some examples. Let's assume I want to say every book's title has been seen by at least one individual.

$$\forall x \exists y | R(x, y)$$

$\forall$ Reads as "every."

x Refers to "book's title."

$\exists$ States as "there exist."

y Refers to "a person."

| Refers to "such that."

$R(x, y)$ Refers to a relation that states the book's title "x" has been seen by person "y."

Well, what we are claiming is clear. You may think that any written book has an author. Therefore, at least the author must have seen the book title. However, if you think further, you may also bring up the books written by blind people that have never been seen by anyone else.

What I would like to get out of this argument is there are many other possibilities in the real world that we most likely need to consider. The other issue is any phenomena we encounter in an intertwined universe can be infinitely vast and unmanageable. Just admitting the complexity is a must towards getting closer to the truth.

The second-order logic is a continuation of the first order logic. A quick example would be every flower on earth either came from the Netherlands or another country. The sentence can be written as follows in logic terms.

$$\forall F \in E | (F \in N \vee F \notin N)$$

$\forall$ Reads as "every."

F Refers to "Flowers."

$\in$ Reads as "element of."

E Refers to "earth."

| Refers to "such that."

∨ Can be read as "or."

∉ Reads as "not an element of."

Hopefully, the mathematical terms and symbols do not leave any ambiguous point; what about the real world? Well, think of the flowers that exist on the exact border line between Belgium and the Netherlands. The other possibility could be the flowers that are in a part of an ocean or island that is not claimed by any country.

The many-sorted logic puts the universe into sets rather than a single homogeneous blend. For instance, instead of having a single driver's licence to operate any vehicle, we have various licences designated for different categories of vehicles. For example, if someone has a truck licence, the individual may not be able to use the permit to operate a lift truck or crane truck.

On the other hand, in many countries, if you have a truck licence, you can drive cars. "Given you have a truck licence, you are allowed to drive a car." Based on the previous sentence, can we conclude you are capable of driving a car? Given you are a truck driver with 40 years of driving experience who never drove anything but the truck during these years, do you think you can drive an electric car without any assistance or debriefing? Would all truck drivers respond to the same question similarly?

Another issue with the categories in this example is the overlapping problem. We wanted to segregate the licences into different sets, but we see a truck driver can drive a car. What I want to get out of this argument is there are more probabilities that we need to address. The categories are hard to define precisely without an overlap. Moreover, the boundaries are not constant and clear in many situations. The dynamicity of the boundaries in categories prevents an affixed logic with absoluteness.

Another form of logic is the modal logic. Basically, in the modal logic, you change part of a sentence. For instance, the sentence "Mathematics explains whatever happens in the universe perfectly" can be modified to "Mathematics tries to explain whatever happens in the universe perfectly."

Aristotle was mostly concerned with the non-modalized logic. It was a United States philosopher, Clarence Irving Lewis, in 1918 who formally worked on modalities. Lewis' works served as a significant step toward the rising of uncertainty in logic and the beginning of statistics.

What we hope to conclude from this chapter is that mathematics is easy to follow. Also, mathematics can be used in association with statistics. Many sorts of logic are mathematics. Even formal logic uses some form of mathematics. However, the main issue is whether mathematics is the best tool that we have to handle everything.

When we use the term mathematics, we have to be very careful how we define it in this book. The reason is mathematics can have a broad meaning. By mathematics, we are talking about the exact and absolute science that studies the relationships between quantities. You may argue that we should present the given definition for the pure mathematics. You may also reason statistics and probability go to the applied branch of mathematics. I believe you are correct.

The only issue that we are trying to discuss in more detail is, the world in totality does not necessarily follow absolute rules. That is, the pure branch of mathematics may not be the best way to reflect the real world. This fact may seem obvious at first glance; however, the reality is we tend to use pure mathematics in real-world problems. In fact, many of our conclusions have been significantly affected by the absolute vision towards any problem.

Regardless of the validity of my claim concerning the power of mathematics, I firmly believe mathematics is beautiful, which I will try to demonstrate in the next chapter.

Math is Beautiful

When we talk about beauty in mathematics, we can look at beauty from several angles. For example, some mathematicians may provide some proofs that are simply beautiful. I think Euler, the Swiss mathematician, dedicated many such proofs to humanity. One may find a mathematical function that explains a part of nature as gorgeous. On the other hand, one may talk about pure mathematics regardless of the application that may or may not be associated with it. In this chapter, I will try to convince you that mathematics is beautiful.

Speaking of beauty, let's talk about people. In general, one reason we presume a particular person to be beautiful is the facial symmetry of that person. That is, we tend to find someone with a symmetrical face to be attractive. Well, the reason for that stems from survival advantages. People with symmetrical faces happen to have fewer dispositioned genes; therefore, they are healthier. While this fact is true, one hundred percent symmetry in one's face is not normal.

The attractiveness to symmetry is a mathematical concept that is passed on to us. It is in our genes to be attracted to such faces and choose them as an ideal partner.

Averageness is another influential factor for attractiveness. Averageness gives us many clues as to why we humans are so obsessed with looking for formulas in nature. It explains why it is so important to look for repetition.

The name averageness on its own does not require any explanation in association with mathematics, in particular, statistics. The reason for that is average has a well-defined meaning in mathematics, and you are more than welcome to conduct research on it. The way our brain identifies averageness is an amazing process, which we have no intention to discuss in detail in the presented context. Regardless of the details, it is interesting how our brain handles averageness identification without any apparent training or calculation.

Well, you may be sceptical about what we are claiming in relationship to beauty in general and mathematical beauty in particular. The concept of mathematical beauty may seem unconventional, and the connection between mathematics and loveliness may not be clear after all. The reality is that mathematical beauty exists, and it is detected in our brain. Researchers at the University College of London revealed the effect on the brain in an engaging manner. The brain scans show that the same emotional brain centres used to appreciate art were active by beautiful mathematics. That is, an appealing equation or mathematical relation would provoke the beauty detection area of the brain.

Let's get out of the discussion of the human face, not because we cannot discuss mathematical features and proportions in human faces. The reason is that, in this book, we want to touch more on a variety of subjects rather delving deeply into a single one.

The other day I was walking, I noticed a tall building. I calculated the number of the floors by counting the number of vertical windows, and I was able to approximate the height of the building. Isn't that cool? Well thousands of years ago, people found similar triangle ratios, and they could locate the height of mountains using a piece of wood. That is cool! To learn more about the procedure, you can read the book "let's laugh at Trig," in which I explain the process.

I am totally against war; however, from the mathematical standpoint, it is interesting to look at some of the battles in the past, for example, the Battle of Trafalgar (1805). The war was between 27 British ships against 33 French and Spanish ships.

The French-Spanish team lost 22 ships. What was the casualty of the English side? Just guess, 27 perhaps? Let me try to describe the combat for you. If you look at the ships back then, they used to have guns and other destructive machinery on either side of their ships. Typically, the two enemies would line up parallel to each other and start shooting at one other.

Logically, one might expect 27 British ships versus 33 French-Spanish would give the French-Spanish a greater chance of winning. One paramount part of the fleet is the flagship, which signals necessary instructions to every other ship. I assume during the battle the phones did not work due to non-payment, so they had to rely on the flagship more than ever. (Yes, I am trying very hard to be funny.)

Nelson, the British mastermind, did something ingenious. Instead of standing parallel to the enemy he decided to go perpendicular on them.

To do so, he put his fleet in two columns. As a result, one column went to the centre and the other column towards the end of the French-Spanish line.

Figure 5

Therefore, he dissected the enemy's ships into three smaller segments. Consequently, flagships could not signal any other ship in the French-Spanish fleet.

The British plan also made it hard for French-Spanish ships to escape without a fight. Moreover, since Nelson believed Englishmen had greater battle morals, individual ship-to-ship encounters was an advantage for them.

The other benefit of the plan was that Nelson put a lot of emphasis on the rear end of the French-Spanish fleet, which bought a lot of time for Englishmen as the Franco-Spanish front needed a lot of time to help the back end of the line. Getting back to the initial question of, "how many ships did the French-Spanish side destroy from the British side?" The answer is none.

Now we can discuss the danger of Nelson's decision, including the inability of the two front ships (from the two English columns) to response the incoming fire from Franco-Spanish ships as they were trying to break the Franco-Spanish line. Of course, many other things could go wrong, one of which resulted in Nelson's death on the battlefield.

One thing is for sure, Nelson carefully calculated the risk, and he changed the history of the world. Why the world? Because the English victory at the Battle of Trafalgar prevented Napoleon from planning to invade England in the future.

If you would like to know more about the mathematics behind it, you can search for Dr. Dan Teague. You should find great formulas associated with the Battle of Trafalgar. The point is that mathematics works beautifully, especially if your opponent has no knowledge of it. There exist at least two explanations for what Nelson had done.

One would be, what Nelson did was based on pure mathematics, and everything happened as he planned. The alternative would be high probabilistic factors tangled in the situation. More investigation is required to answer the questions, but it is a safe bet to leave the subject while getting advantages of the obtained conclusions.

Let's switch gears and talk about a very well-known constant, pi (π). Pi is a constant value, and it take part in many formulas and phenomena associated with various mathematical disciplines.

We postpone any proof about the fact that π is a constant value in any reference frame. At this point, let's discuss a few interesting facts about π, some of which are quite honestly undergoing a lot of debate.

Take the total length of a river and divide it by the straight distance the river travels from the source to the mouth. This would result in a value that scientists call the bendiness of a river.

In other words, this division would lead to a ratio. The bendiness of a river seems a random factor. In 1996, Steven Herrick showed that the average of the bendiness ratios all over the world was equal to the number π.

Total Length of The River

Straight distance

Figure 6

It seems interesting that a constant that arises from the division of circumference of a circle by its diameter is related to the length of the rivers. In general, the value of pi is related to an abundant number of natural phenomena that physical sciences explain.

Moreover, there are many ways to calculate π. One of such is to calculate pi with a needle and two straight lines. The experiment is called Buffon's needle experiment. The experiment goes as follows. You take a piece of metal or wood and draw parallel lines separated by the same distance as the length of the needle. Then drop the piece of metal or wood. You count the number of drops and divide it by the number of times you could hit the lines. Whatever value you get you would multiply by two. The result is an approximation of the value of π.

I am so in love with π that I would like to write a separate book about it. I just want to point out a quick fact about π, which was first suggested by Euler (1775). He hypothesized that π is a transcendental number, and it was proved by Lindemann (1882). That is, π is not a root of any rational number.

Pi, being a transcendental number, is beside the fact that π is irrational, which Lambert proved in 1761. When a number is irrational, it means we cannot put it as a ratio of two integers. Long story short, since we don't know exactly what π is, it not possible to construct or to measure the circumference of a perfect circle.

In case you are interested to know what transcendental numbers are, a brief explanation would be as follows. The transcendental numbers are a group of numbers that are not the root of any non-zero polynomial with rational coefficients. For example, the number $\sqrt{2}$ is irrational since you cannot express it as a ratio of two integers. However, it is not a transcendental number because it is the root of $x^2 = 2$. You cannot write π as a root of any non-zero polynomial with rational coefficient. Coefficients are constants that multiply by the variables of the polynomial. For example, the coefficient of $5x^3$ is number 5.

The interesting thing about the transcendental number is most of our numbers are transcendental numbers. I will not go any further into the topic, but you are more than welcome to study number theory to learn more about this wonderful branch of mathematics.

When we express the value of pi, we do not know its exact value. Note that, as of October 8, 2014; a computer has calculated π to 13.3 trillion (13,300,000,000,000) digits and still going. Now the question is, do we need to know the exact value of π? In other words, can we apply this many digits to any natural phenomena? We will address this type of question in future chapters. The other question that may come to mind is, given π has exact value in the nature, is it possible that we might need a completely different system to calculate π?

This question will not be addressed in this book due to complexity and lack of validity at the moment. We must bear in mind that pi is critically useful, and we know about its contributions to numerous facets of science. Nevertheless, we are incapable of knowing the exact value of pi.

Flowers are just gorgeous. Everyone I have seen likes some flowers. Have you noticed the number of petals in a flower? They either have 1 (White Calla Lily), 2 (Euphorbia), 3 (Lily), 5 (Wild Rose), 8 (Delphinium), 13 (Cineraria), 21 (Aster), 34 (Pyrethrum), 55 (Daisy) or 89 (Daisy). I would like to write the numbers as follow:

1, 1, 2, 3, 5, 8, 13, 21, 34, 55, 89, …

Do you notice we can obtain any given number in the sequence from the third position onward by adding the two previous numbers? For example, 55 is obtained by addition of 34 and 21.

This sequence of numbers is called the Fibonacci number. Fibonacci was a great mathematician (1170-1250), and the current scientific community named the presented sequence after him. Before I go into detail about the sequence, I believe it is not fair to give him all of the credit. An Indian mathematician, famously known as Pingala, wrote the sequence in 200 B.C., which means Pingala defined the sequence around 1400 years before Fibonacci.

The presented sequence starts with two ones, and then each subsequent number results from the addition of the two previous values. This sequence may not seem that useful at the first glance, yet it is. To better understand the calculation, you may refer to the following.

$1 + 1 = 2$

$1 + 2 = 3$

$2 + 3 = 5$

$3 + 5 = 8$

$5 + 8 = 13$

...

Speaking of the originality of the Fibonacci sequence written in 1202, he considered an ideal growth of rabbit populations. Here's how it goes. You have a pair of bunnies (1), and after a period they become a pair of adult rabbits (1). The adult pair would mate and create another pair of bunnies. In total, you have two pairs of rabbits (2).

During the next stage, the adult rabbits produce another pair of bunnies, while the bunnies from the last stage become an adult pair. Therefore, in total you have two pairs of adult rabbits and a pair of bunnies, which adds up to three pairs (3).

This process would go on; however, remember these are ideal situations. This ideal condition is a necessary assumption, and the whole point of this book is about the reality rather than the ideal situation. Imagine what happens if one of the rabbits does not reproduce or reproduces differently.

This doubt is also true for the case of petals in flowers. The fact is we do have flowers with four petals (fuchsia), which is not in the Fibonacci sequence. We discuss this difference afterward in the book. However, the mentioned points do not change the fact that mathematics is mathematics.

At this stage, we will discuss some more cool mathematical findings. The Pascal triangle was introduced by Pascal (1623-1662), or was it? Again, it is not accurate to make such claim. In fact, Pingala (200 B.C.) presented a somewhat similar formula to the modern method associated with the triangle. Two other mathematicians long before Pascal provided the current recipe of the triangle.

The Persian Mathematician, Al-Karaji (953-1029), and the Indian mathematician, Bhattotpala (10th century), gave us the current formula of the Pascal triangle. In Iran, the triangle is named after Khayyam (1048-1131); in China, it is attributed to Yang Hui (1238-1298); in Germany, Stifel (1487-1567) worked on the triangle; and, in Italy, it is named after Tartaglia (1500-1577). All of these scholars obtained the results of Pascal's triangle long before Pascal did.

It is absorbing to find out why the modern scientific community gave most of the credits to people who were not the original discoverers. The mentioned is another beauty of mathematics that shows how important the naming of scientific findings is.

The naming is significant enough that the scientific communities, which are recognised as truth seekers, have accepted the inaccurate information over the truth behind them.

Let's not get into the politics, yet it seems impossible. I wish there were a choice to get into or prevent anything from becoming contaminated with the politics.

Let's take a look at the multi-named triangle and see what is unique about its construction.

```
                    1

              1        1

           1     2     1

        1     3     3     1

     1     4     6     4     1

  1     5    10    10     5     1

1     6    15    20    15     6     1
```

...

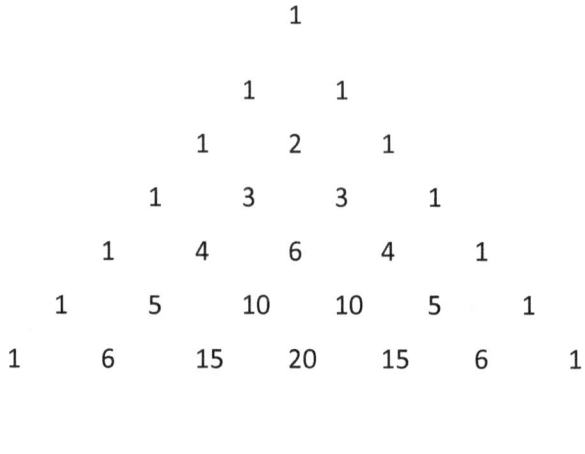

Well, can you detect any pattern? First of all, have you noticed the ones on the sides? Perfect. The numbers in the middle are resulted from the addition of the two numbers immediately on the top. Good to know, now the point is for what purposes can we use this triangle? We can interpret many interesting applications from the triangle.

To give you a few examples, we can name the following. Sierpinski Gasket, Triangular number, tetrahedral numbers, probability problems and many more applications are among the expenditures associated with the triangle. Let's explain a portion of such applications.

At this point, let's discuss multiplication and its relationship to the presented triangle. Let me ask you a few questions. What is the answer to the following questions?

$$(x + y)^0 = 1$$

(The reason for that is any number to the power of 0 is 1.)

 To investigate why you can follow the presented steps and see if it makes sense.

Let assume k is a number and we want to solve

$$\frac{k^{10}}{k^{10}} = 1$$

Or we could use the power rules and obtain the result as follow

$$\frac{k^{10}}{k^{10}} = k^{10-10} = k^0$$

Therefore, we can conclude the following.

$$=> k^0 = 1$$

That is, we can conclude any number to the power of zero is 1.

$$(x + y)^1 = 1x + 1y$$

(The term itself)

$$(x + y)^2 = 1x^2 + 2xy + 1y^2$$

(Multiply $(x + y)$ by itself)

$$(x + y)^3 = 1x^3 + 3x^2y + 3xy^2 + 1y^3$$

(Multiply $(x + y)^2$ by $(x + y)$)

$$(x + y)^4 = 1x^4 + 4x^3y + 6x^2y^2 + 4xy^3 + 1y^4$$

(Multiply $(x + y)^3$ by $(x + y)$) ...

Now let me write the solutions in the following format:

$$1$$

$$1x + 1y$$

$$1x^2 + 2xy + 1y^2$$

$$1x^3 + 3x^2y + 3xy^2 + 1y^3$$

$$1x^4 + 4x^3y + 6x^2y^2 + 4xy^3 + 1y^4$$

...

Look at the coefficients (the numerical values that are in the base). Do you notice they resemble the triangle we revealed earlier? Perfect :D.

Also, if you number the rows in an ascending arrangement from the top to the bottom starting from zero, you notice another interesting relationship. The total value of the exponent of each term is equal to the row values. The row values are also equal to the exponent outside the brackets. Furthermore, note the exponents of the first variable are descending, while the second variable is ascending from zero in each term.

That is, take the row number 4 (the fifth row from the top as you start counting from 0). You have the following:

$$(x + y)^4 = 1x^4 + 4x^3y + 6x^2y^2 + 4xy^3 + y^4$$

The first term is $1x^4$, and the total power of variable is 4.

The second term is $4x^3y$, and the total power of variables are $3 + 1 = 4$.

The third term is $6x^2y^2$, and the total power of variables are $2 + 2 = 4$.

The fourth term is $4xy^3$, and the total power of variables are $1 + 3 = 4$.

The fifth term is y^4, and the total power of variable is 4.

Note on the left-hand side; we have $(x + y)$ raised to the power of 4.

As you can see, everything is equal to 4 as we expected.

Before we escape from the topic, let's quickly discuss a possibly interesting thing. I hope what I am about to say does not change when you try it out while reading this book. If you go to google.com and type in $(x + y)^2$, it graphs the function, which is gorgeous. Now let's say we want to test the maximum number of digits that Google's graphing calculator can take. To demonstrate that, let's put the following equation into the calculator:

$$(x + y)\wedge 3$$

Did you notice the difference between the two graphs? The first one would have a shape with two sides going up, and the second one has one side going up while the other one side goes down. If you try more positive integers (whole numbers), you see the proceeding pattern. For all the even powers, both sides would go up, and for all the odd exponents, the sides would have opposite directions. Now try the following:

$$(x + y)\text{^}1000000000000003$$

The number 1000000000000003 is an odd number, so we expect that one side goes up while the other goes down. However, we see the Google's graphing calculator treat this as an even number. That is, the Google graphing calculator only reads, at most, 14 digits, and it was unable to read the 15th digit.

We did not mention this point to say that Google's calculator is weak. In fact, a 14-digit calculator is very powerful. The point we try to make is, do we need more digits in the real-world situation? For example, do we need to approximate π up to more than 13 trillion digits? The answer to this question would be postpone to later chapters.

Let's get back to our triangle and make another point about it. This time, we draw some tilted lines and add up all the numbers along the line as presented in the following:

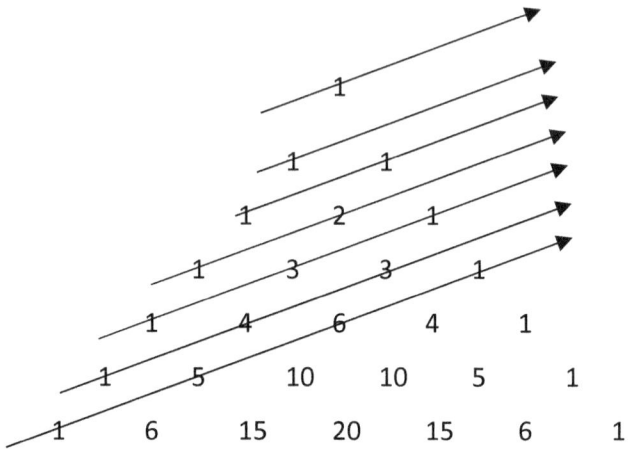

The number on the first line adds up to 1.

The number on the second line adds up to 1 as well.

The numbers on the third line add up to 2 ($1 + 1 = 2$).

The numbers on the fourth line add up to 3 ($1 + 2 = 3$).

The numbers on the fifth line add up to 5 ($1 + 3 + 1 = 5$).

The numbers on the sixth line add up to 8 ($1 + 4 + 3 = 8$).

The numbers on the seventh line add up to 13 ($1 + 5 + 6 + 1 = 13$).

Let's put them in a single row and see if we recognize them.

$1, 1, 2, 3, 5, 8, 13 \; ...$

They look like Fibonacci sequence to me.

That is interesting. Remember addition? And then a quick way to add up is multiplication. A quick way to multiply is using exponent (power). Now we can multiply variables while we employ the properties of the triangle to find the coefficients and the exponents of the variables. The triangle reveals the exponent (a by-product of multiplication) relationship. Can we see a path from exponent to addition?

Is there a formula that we can follow to find the coefficient values without the use of the triangle? Yes, there is a formula, and the Indian mathematician, Mahavira, suggested it in 850. The procedure is called the binomial coefficient formula. I assume the reason we called it binomial has something to do with the fact that there are two variables to the power of a constant. You are more than welcome to read more about it.

Mathematics has a lot of applications, and it gave rise to many branches of science or elevates them to another level. Physics and many engineering branches (if not all) are obvious examples of mathematical applications. However, later on, we see the new findings show direction towards others paths and open new doors to actuality.

Let's take another avenue. One rather interesting thing is an attempt to answer why the average human body temperature is 37°C or 98.6°F, except for my mom, who has an average body temperature of 36.5°C. The answer may lie in the fact that fungal infection happens at a lower temperature.

Casadevall, the professor and chair of microbiology and immunology at the Albert Einstein College of Medicine of Yeshiva, showed that, for every Celsius degree increase in temperature, the fungal infection decreases by 6%. This relationship explains why cold-blooded animals such as lizards are more prone to get fungal infections.

After performing a mathematical optimization, the researchers revealed that the optimal temperature should be close to 36.7°C. Interestingly enough, my mom seems to be closer to the optimized value. At this temperature, there is a correct balance between human susceptibility to fungal infection and not needing to eat constantly to maintain metabolism.

Well, thanks for confirming nature is doing its job. The magnificence of the calculation is that math and nature are in agreement to a great scale. Let's be frank; it is hard for me to write something about the beauty of mathematics. The main reason is, anything I look at is mathematics-related and, therefore, beautiful.

Let's briefly talk about how vision works for a human. Only close to 1% of the information perceived by photoreceptor cells transfers to ganglion cells. Various parts of the brain process the obtained information from the ganglion cells and generate a relevant response to the stimuli. The reason for such selective processing is that the huge volume of data would require an excessive processing time.

The general notion is the occipital lobe, at the back of our brain, is in charge of vision. However, the brain processes information in a more complicated manner. A stream of applied mathematics in neuroscience handles the calculation of the processing times along with various useful ratios. As expected, the mentioned part of mathematics is not intact from the probabilistic calculations. That is, the complexity of the brain does not allow a simple mathematical formula to unlock its secrets. Needless to say, such a simplistic prospect on the brain led to great findings and even a new exciting branch in neuroscience, neuroethology. I encourage you to read more about this wonderful topic.

We can also talk about how long it takes for a glass of hot water to reach room temperature. Of course, we can use Newton's Law of Cooling Formula to calculate it with great precision. There are numerous examples of mathematical application in physics, which I do not intend to discuss at the moment.

One topic I would like to discuss is the optics and a few fundamental laws associated with it, such as the rules that are discoursed by Newton in the 17th century. By the way, are they discovered and explained by Newton for the first time? Ibn Sahl, the Persian mathematician, wrote "on burning mirrors and lenses" in 981. One of his impressive works was driving the "Snell's law of optics" almost 600 years before Willebrord Snellius. Of course, he wrote it first, yet again it is named after Snellius.

The father of optics and one of the first theoretical physicists, Alhazen (Ibn al-Haytham), was an Arab scholar who lived from 965 to 1040. He had great contributions in many disciplines, including optics. The "Book of Optics" is one of his works dedicated to optics in which he explained optics in seven volumes.

Long before Snellius and Newton, other scientists came up with many beautiful laws related to light, lenses, mirrors, and other associated contexts. At this point, we do not have any intention of discussing optics, but we will explain the power of mathematics in the modern era.

One reason we are mentioning the original discoverer's name is to show the importance of the discoverers and designations over the course of history. One of the glories of the winner in our time is the scientific advancement, which should explain why most scientific communities in developed countries are not supportive of true originators from other regions. In a similar manner, during ancient times, the winners would write history as they wanted it to be.

For example, in most cases, the conquerors would not explain the terrible unfair methods they used to win. Also, the winners would exaggerate the power of the enemy to show their incredible power. As I explained, nowadays science is the conqueror. On the account of science's vital power from many prospects, the conquerors have defined the truth differently. The winner writes it all.

About recent stories, in particular, optics, we realize a scientist who did not necessarily discover a critical phenomenon is getting all the credit for it simply because the mentioned scientists are on the winner's side. Although ignorance is not a very truth-seeking act, as it should be in the scientific communities, it has become a fashionable trend.

All stated arguments aside, mathematics on its own is unbelievably beautiful, and the power of it is well known. The attractiveness and influence of mathematics explain why dominant forces seek mathematical knowledge and want to get credit for it.

Math Is Too Simple

We barely touched on the mathematical beauty topic, and I am sure you can find tons of related examples of mathematical magnificence in other sources. Nevertheless, we need to advance an important question. Does mathematics explain reality precisely as it is, or is it just a simplified model? The mentioned issue is the primary purpose of the book.

What we hopefully have gathered from the previous chapters is that mathematics has served us well. That is, mathematics helped humanity both in practical and pure prospects. I am sure you can find a lot of useful resources that would verify my claim.

There is no doubt in my mind that mathematics is a relatively close reflection of the real world. However, as humans further evolve, is math still understandable and applicable? In other words, would mathematics precisely follow nature? What we are going to discuss is the application of mathematics in the real-world situation while we leave pure mathematics and its joyful concepts on the side.

You may argue that even pure mathematics is a reflection of reality as it is real. At least it somewhat resembles the complexity of the human brain. From that sense, you are correct, and that's a topic for another book.

Let me ask you a simple question. Imagine you are a farmer, and you have a box of apples in your storage from yesterday. Today you add another box. How many apple boxes do you expect to have tomorrow?

$1 + 1 = 2$, right? Simple? Not necessarily. Tomorrow you check the storage, and here is what you see in the room. Half of one of the boxes is gone. Probably a group of rodents took the missing apples. The other alternative could have been you open the storage, and you find three boxes since your friend added one box after you left the warehouse yesterday.

Yet another possibility is you open the storage, and you find exactly two boxes. Now the question is, would $1 + 1$ be equal to 1.5, 3, 2 or some other values?

Well, mathematically, $1 + 1 = 2$; however, there are a few points to be considered. You must be knowledgeable about every single thing that might be related directly or indirectly to the part of the event you want to formulate. This task is incredibly difficult, if even possible. In our simple example, if you sit by the boxes all night to make sure every possible incident is under control, a distraction may happen, and you eat one of the apples yourself. It is also possible an earthquake breaks everything, and you lose whatever you had in the first place.

Now the question is, how we can correctly formulate something a lot more complex? How can we make sure about it? I believe at least one of the following is true. We are not capable of precisely watching over everything. The other possibility is that the reality is the world does not follow our mathematical equations. The last mentioned possibility is a topic that we will elaborate more in the coming chapters.

At this point, I would like to draw your attention towards various possibilities that even a simple event can possess. In other words, you need an isolated system if you want to perform even a modest mathematical operation, that is, if you want to add any possibility to your calculations. Now the question is, how often do you think that insulated events occur in the real world both organically and artificially?

Let's not even touch on organic isolation as it appears to be more complex and intertwined in comparison to artificial isolation. You may argue the entire world as a system is an isolated system. You are correct, yet the problem with such a claim is the obstacle in performing the calculations. The only small issue is we are not able to take care of all possible variables in our equation. Let me rephrase. By isolation, we mean a system that has only a few variables so that we can perform a simple mathematical operation on it. In our example above, I put the boxes of apples in storage, and there is no other factor. Therefore, I get two boxes the following day.

With regards to artificial isolation systems, most of the time we are very successful in achieving the goals of experimentation. Nevertheless, there are two problems with the artificial systems. Number one is the results are not always correct and supportive of our expectations. The other issue is the systems are artificial and not a part of the real world. You may again argue if the fake setups exist, then they must be a part of the reality. You are right; however, such models do not necessarily reflect the piece of information about the world that you are trying to comprehend or describe.

Let's get back to $1 + 1$. We did discuss various possibilities and isolation aspects. Now let's discuss another necessary component to perform a seemingly simple addition operation. Starting with a simple question, what is the result of the addition of an apple and an orange? Would it be two apples and two oranges or an apple and an orange?

What I would like to emphasise is, to add two things they must be equal. Now the question is how you can make identical components. If you add an apple to another apple, you have two apples, or do you?

Well not precisely. You need the two apples to be exactly identical. I totally understand I am getting very picky, yet this is the whole point. Now the problem is how to find two identical apples. Let's say we cannot pinpoint two identical apples. Why don't we go to more elementary particles? The reason for this selection would be apparent in a few lines.

When we talk about elementary particles, we are referring to building blocks of everything. In other words, these particles are not made up of anything. One such elementary particle is an electron. Just to clarify, protons are not elementary particles as they are made up of other elementary particles known as quarks, according to the latest findings and theories. I have no intention of delving into particle physics concepts, yet I would like to make a clarification.

Let's say I want to compare two electrons. Well, let's even go one step further, compare an electron with itself. Experimental physicists take an electron and shoot it on a screen. The process is more complicated than what I have explained, but I hope I can convey the point. The electron hits the screen at point A.

Then the physicists would take the same electron, and they shoot it back to the screen with everything kept the same. By everything, we refer to the projection angle, distance to the screen and all other criteria. However, this time, the electron hits the screen at a different point, B. That is, an electron may behave differently under the same situation. We will not discuss the further bizarre things that quantum mechanics explains, yet you are more than welcome to do your research on the topic.

We can conclude at least two facts are related to our argument. First, not even the electrons are identical. The similarity of two things is extremely crucial if we require identical "things" to perform an addition operation. The other issue is randomness factors, to be more precise, the uncertainty principle. I would love to talk about the uncertainty principle in the physical sense; however, that requires a separate section.

In turn, I would like to draw a conclusion from the uncertainty principle. That is, not being one hundred percent sure about any event. The mentioned principle does not work correctly with most of the absolute mathematical prospects. Remember we are investigating addition, a very simple (if not the simplest) mathematical operation. As a result of the quick analysis, we concluded even the addition could be a challenge.

Therefore, it is not 100% correct to even say $1 + 1 = 2$ for a few reasons. There are no two identical elements to add. To be more precise, an element is not equal to itself if we consider the time factor. Again, we are discussing the simplest particles. At this point, we will not address a complex system such as a human being, and how you are different from two sentences before.

The reason we don't consider the complex systems at the time being is that we have failed to show the identity of the simpler models. If the building blocks are different, then it makes sense the more advanced system has a great chance to be different as well. The other critical factor is the countless number of variables to a single event, which we refer to as randomness.

Let's talk about one of my favorite subjects of all times, psychology. As beautiful as it seems, psychology as an independent discipline is relatively new. In the 1870s, it branched off from philosophy and opened its path independently in the scientific community. Psychology soon became a popular subject, and it diverged into other incredible disciplines.

You may ask why independence of psychology is a big deal. The answer can be the independence from philosophy is not our primary concern. Rather, it is the concept of initiating the first experimental laboratory for psychological purposes in 1879 by Wundt in Germany. It is worth mentioning that Wilhelm Wundt is the first person who called himself a psychologist. In that sense, what did prevent scientists from establishing the first laboratory to study psychological effects earlier than 1879?

At this juncture, something interesting shows up. We humans have conducted many experiments on the physical world around us for centuries, if not millennia, before. By the physical world, I am referring to the phenomena handled by physicists. As the physical world contains comparably simpler systems, our simplified model did work to a great extent.

As an example, let's assume we want to calculate the velocity of an object as it hits the ground. The object drops from the Great Trango cliff at a height of 1,340 meters. By using the following formulas, we can find the velocity with a relatively reasonable approximation:

$$mgh = \frac{1}{2}mv^2 =>$$

Where m is the mass of the object,

"g" is gravitational constant, which is not a constant as it depends on many factors, including height. However, we can ignore the mentioned for the sake of the argument and still get a reasonable response.

"h" is the height of the object from a set point.

"v" is the final velocity of the object.

Please be cautious; we are assuming the object is not massless, and it is not losing mass during its journey. In addition, we ignore the air friction and the gravitational force from the cliff as the object falls. We also ignore the gravitational force of the moon and other celestial bodies, which affect the dropped object. Of course, we have to close our eyes to many other forces and relativistic effects.

Then we can continue the formula as follows and still obtain a reasonable result.

Divide both sides of the equation by m (mass).

$$gh = \frac{1}{2}v^2 => (\text{Assuming } g = 9.8\frac{m}{s^2})$$

$$v = -\sqrt{2 \times 9.8 \times 1340} \approx -162.1\frac{m}{s}$$

The value of "v" indicates the speed of the object is almost 162.1 meters per second, and the direction of the movement is towards the ground as I assumed positive to be upwards.

Let's briefly discuss when we began determining the related formulas. It was 1021 when Al-Biruni, the Persian scientist, defined acceleration relation to the non-uniform motion. Al-Biruni discovered friction, and he noted that the centre of earth attracts everything (gravity, which Newton observed by the falling apple over 600 years later). Alhazen, the Arabian Scientist, proposed the concept of inertia between 1000 and 1030, which is known as the first law of Newton. Hence, the first law of Newton is close to 600 years older than Newton.

Avempace, the Andalusian scientist, developed the reaction force concept between the years 1100 and 1138. This law is now famously known as the third law of Newton, which Avempace discovered more than 500 years before Newton was born. It was Al-Baghdadi, the Arabian scientist, who discovered the proportionality between force and acceleration between 1100 and 1165, which is now famously known as the second law of Newton. Al-Baghdadi explored the relationship more than 500 years before Newton.

A very hard-working scientist who was a slave for a period was Al-Khazini, the Persian scientist who explained gravitational potential energy in 1121. The majority of physicists attribute gravitational potential energy to Newtonian mechanics while it was written more than 500 years before Newton's birth. In the 6th century, Philoponus, an Egyptian scientist, explained the projectile motion against gravity. Moreover, it was Al-Bitruji, an Andalusian scientist, who elaborated on gravitational potential energy during the 12th century. The story (history) goes on and on.

Aside from the fact that many of the laws of physics are not reasonably named after the associated discoverer, and the namings can express a lot of politics behind the scientific community. We wanted to present a quick sample of the timeline to show the human advancements in the calculations of a simple natural phenomenon. We have seen formulas did not develop over a few years of a particular scientist's life, but rather over centuries by numerous masterminds. Newton himself beautifully admitted the teamwork of scientists over centuries as he stated in 1676, "If I have seen further, it is by standing on the shoulders of giants."

The timeline shows that physicists had started their experimentation long before psychologists. Please be aware I did not mention the first experiments in physics; rather, I explained the timeline for some of the mechanics laws in physics (the aforementioned free-fall example).

We did not notice a huge problem until the end of the 19th century and the beginning of the 20th century. During this time, a few great theories in physics were raised, namely the relativities and quantum. Bear in mind, we are just talking about the simple part of our physical world, and we noticed the problems after many years.

For one of the most complicated parts of all beings, the human brain, we could not even start on the wrong foot. Of course, the mentioned is my idea, but it might be worth thinking about the reasoning.

Let's say it is the year 1016, and you are a physician with excellent medical and mathematical skills. You have a patient who is raped in the war. Your patient is feeling depressed. If you are asking whether we knew about depression back then, yes, we knew what depression was as the Ebers papyrus in 1550 B.C. discusses clinical depression.

A few days later, you have another patient with the same unfortunate incident as the former patient. However, the problem with the second patient is restricted to severe pain. The second patient gets back to normal life in a few weeks, while the first patient is showing more difficulties as time passes. Can you now talk about the effect that rape has on perception and behaviour of a victim?

Apparently, we are faced with something a lot more complex than the velocity of a ball as it hits the ground when dropped from a cliff. Now the problem is the logic does not work in the simplified format that we foresee in other philosophical matters. In other words, it is not simple mathematics that plugging in the values of the limited number of variables can solve the problem.

In the rape example, if sexual assault then depression or if sexual assault then only pain does not stand. In other words, as I call it, linear logic does not hold for psychological phenomena. The reason for that is we have a lot more variables that we need to consider.

The mentioned issue seemed to be a significant obstacle for a long time. Specifically, the formulation seemed impossible. Does it mean we did not use psychology before the 1870s? We certainly did use psychology as a science to a great extent. We did not formulate many effects and behaviours, but we start our experimentation from the time we are born. We initially collect and store the result of our interaction with the surroundings, so we become experts in using them. The better we employ the results of these experiments, the more maturely we would act in society.

All I want to get out of this discussion is, historically, we lacked the tools to unlock psychology. Mathematics has served us well; however, when we discuss a complex system, mathematics can no longer formulate with one hundred percent accuracy.

Let's get back to the question I asked earlier in the chapter. Why don't we go after complex systems to compare whether they are identical or different? As stated, one reason is that they have different compartments. Therefore, even though the whole picture of two systems might seem very much alike, there is an excellent chance the wholes are not identical as well.

The other reason is that small problems tend to grow in more complex systems. That is, if you and your friend are taking a straight path from a single point at an angle of 0.5 degrees of one another, after a meter you are about 0.008726 meters apart, or almost at the same position. However, after 1,000 kilometers, your distance is close to 8,726 meters. Note this distance is a lot, and it is hard to ignore. You can refer to the following graph to better understand the point.

Figure 7

What we are trying to extract from the given graph is, as we stretch each end of the lines, the distance between the two ends increases. Conversely, in the beginning, the distance between the two lines is almost zero. The mentioned is what we mean by stating the problem with the complex system and how the error grows as well.

You may argue the difference has increased in proportion to the real value. Your argument is valid, yet we have to consider another point. In most cases, we were unable to detect or measure the error as it was comparably small. Nonetheless, later on, when the error gets larger, it becomes detectable. There are many other cases to elaborate further on the point, yet we assume the presented ones are sufficient enough to convey the argument.

One might also argue these kinds of errors has been taken care of by uncertainty calculations that we employ in the lab while performing experimentations. Once again, you are pointing to a pronounced aspect, which is the entire purpose of this book. That is, mathematics is not a complete replica of nature, and some adjustments are needed to make it work. Such changes have revealed themselves during practical calculations of various applications, and, nowadays, we take them for granted. In the following chapter, we will further elaborate on the mentioned topics.

Let's rephrase the conclusion by saying that, although mathematics is a beautiful tool from many senses, it should be used with caution. To be precise, mathematics has a perfect tone in most of its disciplines; however, the reality is that probability happens to be a more realistic model of the world.

I believe the absolute side of mathematics would change our view and behaviour towards every aspect of reality. It might be the case that absoluteness exists for some cases. However, there are discrete spectrums of discontinuity between the absoluteness that we need to investigate with greater skepticism and humbleness. This way we can assume nothing to be true or false in an absolute sense. There are some degrees of possibility for the occurrence of truthfulness or falseness of any event. This prospect leads to an entirely new branch, namely, statistics, which we will explore in more detail in the subsequent chapters.

Statistics Explains

It seems necessary to reveal a bit of history associated with statistics. Statistics is a very new branch of science that started in the 18th century. The term statistics can translate to the science of state. It was originally constructed to assist governments to facilitate administrative and economic tasks better.

That is to say; statistics emerged out of a need to collect and manage big data. From this sense, it should be understandable why statistics is important. First, as we just mentioned, there was a need to manage big data, which previous branches of science or, more specifically, mathematics could not handle it. This essence led to a few hints towards how to deal with big data with lots of variables. Moreover, statistics brought significant outcomes that empowered us to obtain close enough approximations and achieve reasonable conclusions.

The employment of statistics does not mean we humans are getting lazy about going after approximation rather than finding the exact result. Conversely, it can imply there is simply no precise answer to any single question unless a theoretically isolated bound traps the issue.

From the sense that no possible accurate solution exists, we can draw many conclusions. However, first, we need to explore a debate between statistics and mathematics. We need to look at the achievements of mathematics during its kingdom over millennia. Also, we have to look at how statistics has conquered various disciplines during its independence from mathematics.

Regardless of the conclusion, we need to keep in mind that mathematics has at least two sides, one originated to serve humans for a better living experience. The other is comparable to a game for the lover of mathematics. As explained before, mathematicians get pleasure not only by solving problems but also by trying to resolve the issues.

Mathematicians look at a curve, and their dopamine pathway gets activated. Well, this is not just a motivational sentence, yet there has been a study done by the University of Tubingen researchers in Germany to show it. Neuroscientists at Tubingen revealed that dopamine helps people in solving challenging cognitive tasks. Without going very deep into the neurological aspects of the subject, I would like to point out that dopamine is a neurotransmitter associated with happiness, pleasure, and contentment.

Mainly you need dopamine to do serious mathematics, and most likely mathematicians work on related problems to get pleasure out of them. In the battle between statistics and mathematics we introduced earlier, we are not going to discuss the gaming aspect of mathematics for two main reasons. First, I will be highly biased towards the mathematical side as I totally love mathematics even though the limit of my mathematical knowledge approaches zero. The other reason is mathematics had a greater opportunity to evolve.

The previous point does not prevent statistics from being playful. In fact, games associated with uncertainty, probability and, ultimately, statistics are more widely played and accepted. In fact, only math lovers understand and enjoy mathematical games, yet the general public enjoys playing the statistical games. If you have the smallest doubt about it, you can consider the casinos as perfect examples and notice the crowds in them. Additionally, you may ask your surrounding people if they buy lottery tickets. Even other less prominent games, such as soccer with 3.5 billion estimated fans and cricket and hockey with 2.5 billion and 2 billion estimated fans, respectively, were founded on statistics. The probabilistic elements in the mentioned sports make them beautiful and attractive.

In fact, the less probability is involved, and absoluteness dominates, the less widespread a game becomes. One reason could be that improbability seems to be more real, and doubtfulness is what we are accustomed to in comparison to certainty. It is the beauty of soccer that a team with the greatest players does not necessarily get first place in a particular tournament.

Wait a moment, let me think of something else that involves probability. I cannot think of anything that does not have probability. Let's try to start from this moment. There was a chance that I would not have given the example of this instant. In that case, my example would involve something else. However, the ultimate result would have been somewhat similar. I wanted to provide an example to explain the usefulness of statistics.

The point is you could not guess for sure what my next example is. Let me tell you a story of the Second World War. Abraham Wald was a great mathematician who had a rather interesting life. I do encourage you to read about him in your spare time. While he was in charge of looking over airplane damages caused by German forces, he noticed some interesting trends. He found the fuel system of returned plans sustained more damage in comparison to their engines. Now these statistics helped him to make an unusual decision to save many lives. What would you do had you received such information?

He focused on protecting the fuel system. No, he did not. He put the emphasis on the engine system. The reason was he realized that the planes with damage to their fuel system could ultimately make it home. However, the planes with engine damage couldn't even get back.

The point is statistics has served us well during history, and there are a lot of such examples we could present. Other than the importance of collecting data related to an event, it is crucial to understand them correctly. We have no intention of elaborating on data analysis and interpretation at this point, yet you are more than welcome to study more about them at your spare time.

Let's look at the timeline for one of the most revolutionary theories for explaining the nature in the modern era, quantum mechanics. To put things in perspective, the black body concept is an important beginning milestone in the history of quantum physics.

Kirchhoff, a great German physicist, proposed that a perfect black body absorbs all the light. The perfect black body was shown to be more of a theoretical concept rather than a practical model. Nonetheless, it is considered an important step in the history of quantum physics. The black body radiation is only dependent upon the temperature of the object. Moreover, the fact that perfect black body radiation does not exist is a valuable piece of information. It articulates that we need to presume absoluteness with more caution.

The second great milestone in the evolution of quantum physics was the concept of statistical mechanics. Great physicists, such as James Clerk Maxwell from England, Ludwig Boltzmann from Austria, and Willard Gibbs from the United States, introduced statistical mechanics. I have no intention of explaining what they did as it requires a separate book to present some prerequisite knowledge beforehand. However, what we can perceive is the name statistical mechanics suggests it uses statistics to define the average behaviour of the system.

Accepting that there is an average response even for a well-behaved system, is considered a breakthrough. Note the average may cover a vast range of likelihoods; however, we should be cautious that it is not expected to cover every single phenomenon. Even if all the previous trials follow an absolute response, we cannot discuss the future with absolute certainty. Time is a factor that needs to be considered if we are after the absoluteness, and to talk about absoluteness we need to lock time in its place.

Locking time seems to be unconventional. According to the current physical findings, any dynamic system involves the passing of time. Namely, if we stop time, the first consequence would be a static picture.

The other consequence can be a dynamic system, which seems a bit complicated. To lock time, we are required to move at the speed of light.

At such a speed, if we move in any direction within our frame, which moves with the speed of light, we can claim that we made a movement at the locked time. In such case, our speed would become faster than the speed of light. Such speed, according to our current findings and theories, seems a bit unconventional.

To learn more about the reasoning behind it, you are more than welcome to investigate further the special theory of relativity, which is an amazingly beautiful set of rules suggested by Albert Einstein, the German physicist.

Now let's consider how much we can divide something and still get an applicable result. It is correct that we can find the number pi up to billions of digits using an infinite series. One such beautiful series is known as Leibniz series, even though Madhava of Sangamagrama, the Indian mathematician, discovered the formula almost 300 years before Leibniz from the Roman Empire. Not to mention, Madhava's formula was a lot more advanced than Leibniz's rediscovery.

However, the question is, would it give us any benefit regarding applying it to a physical phenomenon? Although the joyfulness in the calculation of π is undeniable, we cannot use the obtained value with such precision in experimentation.

According to a clay tablet found in Babylon and the Rhind papyrus found in Egypt, the obsession with accurately finding number π started close to 4,000 years ago. Back then, we knew the value of π within 1% of the true value.

In case you want to know how reliable 1% is, please consider reading about Z540.3's 2% rule used by the National Aeronautics and Space Administration (NASA) to standardize many operations and equipment. Just to clarify, within 1% of true value is more accurate than within 2% of the true amount.

What we would like to get from this comparison is not saying having π in billions of digits is wrong. Rather, so many digits may not have any practical benefit in real-world applications. The presented example for the value of π serves as a sample that maybe an attitude towards finding absolute accuracy is not the way we should view the world.

Even though I have a great desire to learn and talk about physical sciences, I would like to change gears and quickly introduce another subject.

Let's find out what kind of car you like the most. I hope you have an answer in your mind. The answer can be in any form. It may be something like, "I don't have a favorite car", or the answer can be a question, "from what sense you define the best car?" All of the responses serve the point in a sense that makes you think about a reply. Let's talk about thinking. Now the question is, from where do you retrieve the answers?

Every second, there is an enormous amount of information rushing into the brain. There are close to 100 billion neurons in the brain, and each neuron is accompanied by ten glial cells. Each neuron signals five to 10 impulses per second. Moreover, other than performing tasks, such as nourishing the neurons, the glial cells might be in charge of some signaling as well. That is, a lot of serious talking happens in the brain.

What I would like to get into is we are talking about an enormous amount of input that rushes into the brain. This incoming information would reveal as neuro-firings. The firing can be excitatory or inhibitory. What should mainly concern us at this point is whether the brain processes everything that it perceives. Again, we are talking about millions of inputs per second. Most neuroscientists and psychologists believe the brain selectively processes a small fraction of the incoming data within the conscious mind.

That is, most of the information is either not processed and lost, or processed and not accessible in the conscious brain. According to what we know today about the information in the brain, a significant amount of the data are processed and stored in the unconscious brain.

The unconsciousness is an interesting concept that initially ancient cultures, such as Hindu, introduced. Sigmund Freud, from the Austrian Empire, was not the first person to announce and work on unconsciousness in the contemporary description; however, he helped massively publicize the unconsciousness.

Without going into details of the unconscious, subconscious and conscious, although fantastic to know, we proceed to one of the findings in psychology. As the information is stored and processed at the unconscious level, there are specific ways to retrieve the information or merely use them.

Quite honestly, the declared methods are widely employed in various industries, including adverting and marketing. Indeed, one reason I selected the title of "MATH IS BAD" for the presented book was to use a similar effect. It might be interesting to know about a study conducted by Beilock from the psychology department of the University of Chicago.

Beilock found that "during anticipation period, and not the actual doing of the mathematics," some areas that are associated with pain are activated. That means, not doing mathematics yet thinking about doing mathematics can induce pain in the brain. As a result, anticipating doing mathematics may lead to anxiety. Well, I already knew about the perception of most people over mathematics and the unconscious mind. On the other hand, there was a great chance you did not know about the mentioned research.

Regardless of your knowledge about the issue, most likely your brain would have experienced the pain. The pain information must have been stored somewhere in your brain. I used the possible true information about your unconscious perception to get your attention.

Not to mention, if I say "MATH IS BAD", then I am proposing an absolute sentence, which opposes all I am saying in the presented book. Hence, there could have been another meaning to it. What if by saying "MATH IS BAD", I meant, math is the "**Best Approximation for Determination**" as of now.

Besides, probability and statistics are both by-products of mathematics. Moreover, in the provided text, we are trying to segregate the sense of absoluteness from any phenomena. At the same time, we do not seek to demolish the concept of calculation.

You may argue any sentence that I propose would include a sense of absoluteness, which is entirely correct. However, I am trying to show a gap between absoluteness and reality. This difference is primarily helpful from many prospects. The gap had led to many fascinating findings in several disciplines. As presented in the earlier chapters, a decent example would be psychology.

In psychology, when we talk about probability in a more formal way using calculations, we could make comments about more complex beings with understandable and applicable conclusions. The other examples that we discussed happened to be the modern physics concepts and how statistics plays a crucial role in their foundation.

Indeed, there are a lot more points to argue in association with the presented topic. However, I am a great fan of minimalism. Hopefully, the given text was able to show the viewpoint that we know nothing for sure.

The presented prospect may help us to judge with more caution. As an example, it is true that the Amazon rainforest supplies 20% of our oxygen. Conversely, It does not mean the Sahara is a useless section of Earth simply because it is dry. The Sahara transports close to 40 million tons of mineral each year to the rainforest and makes Amazon a strong lung for our planet. Almost everything is connected, and the intertwined connection makes the universe more complicated than it appears. Therefore, maybe anything is possible.

Along the same line, we try to point out that statistics is a great gift brought by mathematics and philosophy. This gift has led to the blossoming of significant findings, and it has opened numerous doors to a more understandable future with endless possibilities. It may give us reason to bow down to what is in the universe, and be at peace with whatever it contains.

Index

www.ingramcontent.com/pod-product-compliance
Lightning Source LLC
Chambersburg PA
CBHW060636210326
41520CB00010B/1621